Dr. Langhorne

The Fables of Flora

Dr. Langhorne

The Fables of Flora

ISBN/EAN: 9783744785594

Printed in Europe, USA, Canada, Australia, Japan

Cover: Foto ©berggeist007 / pixelio.de

More available books at **www.hansebooks.com**

THE

FABLES

OF

FLORA.

BY

DR. *LANGHORNE.*

..... SYLVAS, SALTUSQUE SEQUAMUR
INTACTOS ..... VIRG.

LONDON:

PRINTED BY T. RICKABY,

FOR E. AND S. HARDING, PALL-MALL.

ADVERTISEMENT.

———

*IN the following Poems, the plan of Fable
is somewhat enlarged, and the province so far
extended, that the original* NARRATIVE *and*
MORAL *may be accompanied with imagery, de-
scription, and sentiment. The scenery is formed
in a department of Nature adapted to the genius
and disposition of* POETRY; *where she finds
new objects, interests, and connexions, to exercise
her fancy and her powers. If the execution,
therefore, be unsuccessful, it is not the fault of
the Plan, but the Poet.*

FABLE I.

THE SUN-FLOWER AND THE IVY.

As duteous to the place of prayer
   Within the convent's lonely walls,
The holy fisters ftill repair,
   What time the rofy morning calls:

So fair, each morn, fo full of grace,
   Within their little garden reared,
The flower of Phoebus turned her face
   To meet the Power fhe loved and feared.

And where, along the rifing fky,
   Her God in brighter glory burned,
Still there her fond obfervant eye,
   And there her golden breaft fhe turned.

When calling from their weary height
   On weftern waves his beams to reft,
Still there fhe fought the parting fight,
   And there fhe turned her golden breaft.

But foon as night's invidious fhade
   Afar his lovely looks had borne,
With folded leaves and drooping head,
   Full fore fhe grieved, as one forlorn.

Such duty in a flower difplayed
  The holy fifters fmiled to fee,
Forgave the Pagan rites it paid,
  And loved its fond idolatry.

But painful ftill, though meant for kind,
  The praife that falls on Envy's ear!
O'er the dim window's arch entwined,
  The canker'd I v y chanced to hear.

And " See," fhe cried, " that fpecious flower,
  " Whofe flattering bofom courts the fun,
" The pageant of a gilded hour,
  " The convent's fimple hearts hath won !

" Obfequious meannefs ! ever prone
  " To watch the patron's turning eye ;
" No will, no motion of its own !
  " 'Tis this they love, for this they figh :

" Go, fplendid fycophant! no more
    " Difplay thy foft feductive arts!
" The flattering clime of courts explore,
    " Nor fpoil the convent's fimple hearts.

" To me their praife more juftly due,
    " Of longer bloom, and happier grace!
" Whom changing months unaltered view,
    " And find them in my fond embrace."

" How well," the modeft flower replied,
    " Can ENVY's wrefted eye elude
" The obvious bounds that ftill divide
    " Foul FLATTERY from fair GRATITUDE.

" My duteous praife each hour I pay,
    " For few the hours that I muft live ;
" And give to him my little day,
    " Whofe grace another day may give.

" When low this golden form ſhall fall
  " And ſpread with duſt its parent plain;
" That duſt ſhall hear his genial call,
  " And riſe, to glory riſe again.

" To thee, my gracious power, to thee
  " My love, my heart, my life are due!
" Thy goodneſs gave that life to be;
  " Thy goodneſs ſhall that life renew.

" Ah me! one moment from thy ſight
  " That thus my truant-eye ſhould ſtray!
" The God of glory ſets in night;
  " His faithful flower has loſt a day."

Sore grieved the flower, and drooped her head;
  And ſudden tears her breaſt bedewed:
Conſenting tears the ſiſters ſhed,
  And, wrapt in holy wonder, viewed.

With joy, with pious pride elate,
  " Behold," the aged abbefs cries,
" An emblem of that happier fate
  " Which heaven to all but us denies.

" Our hearts no fears but duteous fears,
  " No charm but duty's charm can move;
" We fhed no tears but holy tears
  " Of tender penitence and love.

" See there the envious world pourtrayed
  " In that dark look, that creeping pace!
" No flower can bear the Ivy's fhade;
  " No tree fupport its cold embrace.

" The oak that rears it from the ground,
  " And bears its tendrills to the fkies,
" Feels at his heart the rankling wound,
  " And in its poifonous arms he dies."

Her moral thus the matron read,
   Studious to teach her children dear,
And they by love, or duty led,
   With pleasure heard, or seemed to hear.

Yet one less duteous, not less fair,
   (In convents still the tale is known)
The fable heard with silent care,
   But found a moral of her own.

The flower that smiled along the day,
   And drooped in tears at evening's fall;
Too well she found her life display,
   Too well her fatal lot recall.

The treacherous Ivy's gloomy shade,
   That murdered what it most embraced,
Too well that cruel scene conveyed
   Which all her fairer hopes effaced.

Her heart with filent horror fhook ;
   With fighs fhe fought her lonely cell :
To the dim light fhe caft *one* look :
   And bade *once more* the world *farewell.*

FABLE II.

THE EVENING PRIMROSE.

THERE are that love the shades of life,
  And shun the splendid walks of Fame;
There are who hold it rueful strife,
  To risque AMBITION's losing game:

That far from ENVY's lurid eye
   The faireſt fruits of GENIUS rear,
Content to ſee them bloom and die
   In Friendſhip's ſmall but kindly ſphere.

Than vainer flowers tho' ſweeter far,
   The Evening Primroſe ſhuns the day ;
Blooms only to the weſtern ſtar,
   And loves its ſolitary ray.

In EDEN's vale an aged hind,
   At the dim twilight's cloſing hour,
On his time-ſmoothed ſtaff reclined,
   With wonder viewed the opening flower.

" Ill-fated flower, at eve to blow,"
   In pity's ſimple thought he cries,
" Thy boſom muſt not feel the glow
   " Of ſplendid ſuns, or ſmiling ſkies.

" Nor thee, the vagrants of the field,

   " The hamlet's little train behold ;

" Their eyes to sweet oppreffion yield,

   " When thine the falling fhades unfold.

" Nor thee the hafty fhepherd heeds,

   " When love has filled his heart with cares,

" For flowers he rifles all the meads,

   " For waking flowers—but thine forbears.

" Ah ! wafte no more that beauteous bloom

   " On night's chill fhade, that fragrant breath,

" Let fmiling funs thofe gems illume !

   " Fair flower, to live unfeen is death."

Soft as the voice of vernal gales

   That o'er the bending meadow blow,

Or ftreams that fteal thro' even vales,

   And murmur that they move fo flow :

Deep in her unfrequented bower,
　　Sweet Philomela poured her ſtrain ;
The bird of eve approved her flower,
　　And anſwered thus the anxious ſwain.

　　　　Live unſeen!
By moonlight ſhades, in valleys green,
　　Lovely flower, we'll live unſeen.
Of our pleaſures deem not lightly,
Laughing day may look more ſprightly,
　　But I love the modeſt mien,
　　Still I love the modeſt mien
Of gentle evening fair, and her ſtar-trained queen.

　　Didſt thou, ſhepherd, never find,
　　Pleaſure is of penſive kind ?
　　Has thy cottage never known
　　That ſhe loves to live alone ?
　　Doſt thou not at evening hour
　　Feel ſome ſoft and ſecret power,

Gliding o'er thy yielding mind,
Leave fweet ferenity behind ;
While all difarmed, the cares of day
Steal thro' the falling gloom away ?
Love to think thy lot was laid
In this undiftinguifhed fhade.
Far from the world's infectious view,
Thy little virtues fafely blew.~
Go; and in day's more dangerous hour,
Guard thy emblematic flower.

FABLE III.

THE LAUREL AND THE REED.

THE* Reed that once the shepherd blew
  On old CEPHISUS' hallowed side,
   To SYLLA's cruel bow applied,
Its inoffensive master slew.

  * The reeds on the banks of the Cephisus, of which the shepherds made their pipes, Sylla's soldiers used for arrows.

Stay, bloody foldier, ftay thy hand,
   Nor take the fhepherd's gentle breath :
Thy rage let innocence withftand ; .
   Let mufic foothe the thirft of death

He frowned—He bade the arrow fly—
   The arrow fmote the tuneful fwain ;
No more its tone his lip fhall try,
   Nor wake its vocal foul again.

CEPHISUS, from his fedgy urn,
   With woe beheld the fanguine deed :
He mourned, and, as they heard him mourn,
   Affenting fighed each trembling Reed.

" Fair offspring of my waves," he cried ;
   " That bind my brows, my banks adorn,
" Pride of the plains, the rivers' pride,
   " For mufic, peace, and beauty born !

" Ah! what, unheedful have we done !
" What dæmons here in death delight ?
" What fiends that curfe the focial fun ?
" What furies of infernal night ?

" See, fee my peaceful fhepherds bleed !
" Each heart in harmony that vyed,
" Smote by its own melodious Reed,
Lies cold, along my blufhing fide.

" Back to your urn, my waters fly;
" Or find in earth fome fecret way ;
" For horror dims yon confcious fky,
" And hell has iffued into day.""

Thro' DELPHI's holy depth of fhade
The fympathetic forrows ran ;
While in his dim and mournful glade
The genius of her groves began—

" In vain CEPHISUS fighs to fave

   " The fwain that lòves his watry mead,

" And weeps to fee his reddening wave,

   " And mourns for his perverted Reed:

" In vain my violated gròves

   " Muft I with equal grief bewail,

" While defolation fternly roves,

   " And bids the fanguine hand affail.

" God of the genial ftream, behold

   " My laurel fhades of leaves fo bare !

" Thofe leaves no poet's brows enfold,

   " Nor bind APOLLO's golden hair.

" Like thy fair offspring, mifapplied,

   " Far other purpofe they fupply ;

The murderer's burning cheek to hide,

   " And on his frownful temples die.

" Yet deem not thofe of Pluto's race,

   " Whom wounded Nature fues in vain;

" Pluto difclaims the dire difgrace,.

   " And cries, indignant, " They are men."

FABLE IV.

THE GARDEN-ROSE AND THE WILD-ROSE.

As Dee, whofe current free from ftain,
Glides fair o'er Merioneth's plain,
By mountains forced his way to fteer
Along the lake of Pimble Mere,

Darts fwiftly thro' the ftagnant mafs,
His waters trembling as they pafs,
And leads his lucid waves below,
Unmixed, unfullied as they flow—
So clear thro' life's tumultuous tide,
So free could THOUGHT and FANCY glide ;
Could HOPE as fprightly hold her courfe,
As firft fhe left her native fource,
Unfought in her romantic cell
The keeper of her dreams might dwell.

But, ah ! they will not, will not laft—
When life's firft fairy ftage is paft,
The glowing hand of HOPE is cold ;
And FANCY lives not to be old.
Darker, aad darker all before ;
We turn the former profpect o'er ;
And find in MEMORY's faithful eye
Our little ftock of pleafures lie.

Come, then; thy kind receffes ope!
Fair keeper of the dreams of HOPE!
Come with thy vifionary train;
And bring my morning fcenes again!

To ENON's wild and filent fhade,
Where oft my lonely youth was laid;
What time the *woodland* GENIUS came,
And touched me with his holy flame.—

Or, where the hermit, BELA, leads
Her waves thro' folitary meads;
And only feeds the defart-flower,
Where once fhe foothed my flumbering hour:
Or roufed by STANMORE's wintry fky,
She wearies echo with her cry;
And oft, what florms her bofom tear,
Her deeply-wounded banks declare.

Where EDEN's fairer waters flow,
By MILTON's bower, or OSTY's brow,
Or BROCLEY's alder-fhaded cave,
Or, winding round the Druid's grave,
Silently glide, with pious fear
To found his holy flumbers near.—

  To thefe fair fcenes of FANCY's reign,
O MEMORY! bear me once again :
For, when life's varied fcenes are paft,
'Tis fimple Nature charms at laft.

'Twas thus of old a poet prayed ;
  Th' indulgent power his prayer approved,
And, ere the gathered Rofe could fade,
  Reftored him to the fcenes he loved.

A Rofe, the poet's favourite flower,
 From FLORA's cultured walks he bore;
No fairer bloomed in EsHER's bower,
 Nor PRIOR's charming CHLOE wore.

No fairer flowers could FANCY twine
 To hide ANACREON's fnowy hair;
For there ALMERIA's bloom divine,
 And ELLIOT's fweeteft blufh was there.

When fhe, the pride of courts, retires,
 And leaves for fhades, a nation's love,
With awe the village maid admires,   [moves.
 How WALDEGRAVE looks, how WALDEGRAVE

So marvelled much in ENON's fhade
 The flowers that all uncultured grew,
When there the fplendid rofe difplayed
 Her fwelling breaft, and fhining hue.

Yet one, that oft adorned the place,
　　Where nów her gaudy rival reigned,
Of fimpler bloom, but kindred race,
　　The penfive EGLANTINE complained.—

" Miftaken youth," with fighs fhe faid,
　" From nature and from me to ftray!
" The bard, by fplendid forms betrayed,
　" No more fhall frame the purer lay.

" Luxuriant, like the flaunting Rofe,
　" And gay the brilliant ftrains may be,
" But far, in beauty, far from thofe,
　" That flowed to nature and to me."

The poet felt, with fond furprize,
　The truths the fylvan crĭtic told;
And " though this courtly Rofe," he cries,
　" Is gay, is beauteous to behold;

" Yet, lovely flower, I find in thee

   " Wild fweetnefs which no words exprefs,

" And charms in thy fimplicity,

   " That dwell not in the pride of drefs."

F A B L E V.

THE VIOLET AND THE PANSY.

SHEPHERD, if near thy artlefs breaft
 The God of fond defires repair;
Implore him for a gentle gueft,
 Implore him with unwearied prayer.

Should beauty's foul-enchanting fmile,
  Love-kindling looks, and features gay,
Should thefe thy wandering eye beguile,
  And fteal thy warelefs heart away ;

That heart fhall foon with forrow fwell,
  And foon the erring eye deplore,
If in the beauteous bofom dwell
  No gentle virtue's genial ftore.

Far from his hive one fummer-day
  A young and yet unpractifed bee,
Borne on his tender wings away,
  Went forth the flowery world to fee.

The morn, the noon, in play he paffed,
  But when the fhades of evening came,
No parent brought the due repaft,
  And faintnefs feized his little frame.

By nature urged, by inftinct led,
　　The bofom of a flower he fought,
Where ftreams mourned round a moffy bed,
　　' And violets all the bank enwrought.

Of kindred race, but brighter dies,
　　On that fair bank a Panfy grew,
That borrowed from indulgent fkies
　　' A velvet fhade and purple hue.

The tints that ftreamed with gloffy gold,
　　The velvet fhade, the purple hue,.
The ftranger_wondered to behold,
　　And to its beauteous bofom flew.

Not fonder hafte the lover fpeeds,
　　At evening's fall, his fair to meet,
When o'er the hardly-bending meads
　　He fprings on more than mortal feet.

Nor glows his eye with brighter glee,
   When ſtealing near her orient breaſt,
Than felt the fond enamoured bee,
   When firſt the golden bloom he preſt.

Ah! pity much his youth untried,
   His heart in beauty's magic ſpell!
So never paſſion thee betide,
   But where the genial virtues dwell.

In vain he ſeeks thoſe virtues there;
   No ſoul-ſuſtaining charms abound:
No honeyed ſweetneſs to repair
   The languid waſte of life is found.

An aged bee, whoſe labours led
   Thro' thoſe fair ſprings, and meads of gold,
His feeble wing, his drooping head
   Beheld, and pitied to behold.

" Fly, fond adventurer, fly the art
  " That courts thine eye with fair attire ;
" Who fmiles to win the heedlefs heart,
  " Will fmile to fee that heart expire.

" This modeft flower of humbler hue,
  " That boafts no depth of glowing dyes,
" Arrayed in unbefpangled blue,
  " The fimple cloathing of the fkies—

" This flower, with balmy fweetnefs bleft,
  " May yet thy languid life renew :"
He faid, and to the Violet's breaft
  The little vagrant faintly flew.

Gobard   Pub by E & S Harding Pall Mall Oct 30   Burell

FABLE VI.

THE QUEEN OF THE MEADOW AND THE CROWN IMPERIAL.

FROM BACTRIA's vales, where beauty blows
   Luxuriant in the genial ray ;
Where flowers a bolder gem difclofe,
   And deeper drink the golden day :

From BACTRIA's vales to BRITAIN's shore
   What time the CROWN IMPERIAL came,
Full high the stately stranger bore
   The honours of his birth and name.

In all the pomp of eastern state,
   In all the eastern glory gay,
He bade, with native pride elate,
   Each flower of humbler birth obey.

O, that the child unborn might hear,
   Nor hold it strange in distant time,
That freedom even to flowers was dear,
   To flowers that bloomed in Britain's clime!

Thro' purple meads, and spicy gales,
   Where STRYMON's* silver waters play,
While far from hence their goddess dwells,
   She rules with delegated sway.
          * The Ionian Strymon.

That fway the Crown Imperial fought,
  With high demand and haughty mien:
But equal claim a rival brought,
  A rival called the Meadow's Queen.

" In climes of orient glory born,
  " Where beauty firft and empire grew;
" Where firft unfolds the golden morn,
  " Where richer falls the fragrant dew:

" In light's ethereal beauty dreft,
  " Behold," he cried, " the favoured flower,
" Which Flora's high commands inveft
  " With enfigns of imperial power!

" Where proftrate vales, and blufhing meads,
  " And bending mountains own his fway,
" While Persia's lord his empire leads,
  " And bids the trembling world obey;

' While blood bedews the ftraining bow,
  " And conqueft rends the fcattered air,
" 'Tis mine to bind the victor's brow,
  " And reign in envied glory there.

  Then lowly bow, ye Britifh flowers!
  " Confcfs your monarch's mighty fway,
" And own the only glory yours,
  " When fear flies trembling to obey."

He faid, and fudden o'er the plain,
  From flower to flower, a murmur ran,
With modeft air, and milder ftrain,
  When thus the MEADOW's QUEEN began.

" If vain of birth, of glory vain,
  " Or fond to bear a regal name,
" The pride of folly brings difdain,
  " And bids me urge a tyrant's claim:

" If war my peaceful realms affail,

   " And then, unmoved by pity's call,

" I fmile to fee the bleeding vale,

   " Or feel one joy in nature's fall.

" Then may each juftly vengeful flower

   " Purfue her Queen with generous ftrife,

" Nor leave the hand of lawlefs power

   " Such compafs on the fcale of life.

" One fimple virtue all my pride!

   " The wifh that flies to mifery's aid;

" The balm that ftops the crimfon tide*,

   " And heals the wounds that war has made."

* The property of that flower.

Their free confent by Zephyrs borne,
   The flowers their MEADOW's QUEEN obey;
And fairer blufhes crowned the morn,
   And fweeter fragrance filled the day.

FABLE VII.

THE WALL-FLOWER.

" WHY loves my flower, the sweetest flower

" That swells the golden breast of May,

" Thrown rudely o'er this ruined tower,

" To waste her solitary day?

" Why, when the mead, the fpicy vale,
    " The grove and genial garden call,
" Will fhe her fragrant foul exhale,
    " Unheeded on the lonely wall?

" For never fure was beauty born
    " To live in death's deferted fhade!
" Come, lovely flower, my banks adorn,
    " My banks for life and beauty made."

Thus PITY waked the tender thought,
    And by her fweet perfuafion led,
To feize the hermit-flower I fought,
    And bear her from her ftony bed.

I fought—but fudden on mine ear
    A voice in hollow murmurs broke,
And fmote my heart with holy fear—
    The GENIUS *of the Ruin* fpoke.

" From thee be far th' ungentle deed,
  " The honours of the dead to fpoil,
" Or take the fole remaining meed,
  " The flower that crowns their former toil!

" Nor deem that flower the garden's foe,
  " Or fond to grace this barren fhade;
" 'Tis NATURE tells her to beftow
  " Her honours on the lonely dead.

" For this, obedient Zephyrs bear
  " Her light feeds round yon turret's mold,.
" And undifperfed by tempefts, there
  " They rife in vegetable gold.

" Nor fhall thy wonder wake to fee
  " Such defart fcenes diftinction crave;
" Oft have they been, and oft fhall be,
  " Truth's, Honour's, Valour's, Beauty's grave.

" Where longs to fall that rifted fpire,

 " As weary of th' infulting air ;

" The poet's thought, the warrior's fire,

 " The lover's fighs are fleeping there.

" When that too fhakes the trembling ground,

 " Borne down by fome tempeftuous fky,

" And many a flumbering cottage round

 " Startles—how ftill their hearts will lie!

" Of them who, wrapt in earth fo cold,

 " No more the fmiling day fhall view,

" Should many a tender tale be told ;

 " For many a tender thought is due.

" Haft thou not feen fome lover pale,

 " When evening brought the penfive hour,

" Step flowly o'er the fhadowy vale,

 " And ftop to pluck the frequent flower?

" Thofe flowers he furely meant to ftrew
   " On loft affection's lowly cell;
" Tho' there, as fond remembrance grew,
   " Forgotten, from his hand they fell.

" Has not for thee the fragrant thorn
   " Been taught her firft rofe to refign?
" With vain but pious fondnefs borne
   " To deck thy NANCY's honoured fhrine!

' 'Tis NATURE pleading in the breaft,
   " Fair memory of her works to find;
" And when to fate fhe yields the reft,
   " She claims the monumental mind.

" Why, elfe, the o'ergrown paths of time
   " Would thus the lettered fage explore,
" With pain thefe crumbling ruins climb,
   " And on the doubtful fculpture pore?

" Why feeks he with unwearied toil
  " Thro' death's dim walks to urge his way,
" Reclaim his long afferted fpoil,
  " And lead OBLIVION into day?

" 'Tis NATURE prompts, by toil or fear
  " Unmoved, to range thro' death's domain:
" The tender parent loves to hear
  " Her children's ftory told again.

" Treat not with fcorn his thoughtful hours,
  " If haply near thefe haunts he ftray;
" Nor take the fair enlivening flowers
  " That bloom to cheer his lonely way."

FABLE VIII.

THE TULIP AND THE MYRTLE.*

'TWAS on the border of a ſtream
A gaily painted Tulip ſtood,
And, gilded by the morning beam,
Surveyed her beauties in the flood.

* This fable was firſt publiſhed in a Collection of Letters, ſup-
poſed to have paſſed between St. Evremond and Waller.

And fure, more lovely to behold,
   Might nothing meet the wiftful eye,
Than crimfon fading into gold,
   In ftreaks of faireft fymmetry.

The beauteous flower with pride elate,
   Ah me! that pride with beauty dwells!
Vainly affects fuperior ftate,
   And thus in empty fancy fwells.

" O luftre of unrivalled bloom!
   " Fair painting of a hand divine!
" Superior far to mortal doom,
   " The hues of heaven alone are mine!

" Away, ye worthlefs, formlefs race!
   " Ye weeds that boaft the name of flowers!
" No more my native bed difgrace,
   " Unmeet for tribes fo mean as yours!

" Shall the bright daughter of the Sun

  " Affociate with the fhrubs of earth?

" Ye flaves, your fovereign's prefence fhun!

  " Refpect her beauties and her birth.

" And thou, dull, fullen ever-green!

  " Shalt thou my fhining fphere invade?

" My noon-day beauties beam unfeen,

  " Obfcured beneath thy dufky fhade!"

" Deluded flower!" the Myrtle cries,

  " Shall we thy moment's bloom adore?

" The meaneft fhrub that you defpife,

  " The meaneft flower has merit more.

" That daify in its fimple bloom,

  " Shall laft along the changing year:

" Blufh on the fnow of winter's gloom,

  " And bid the fmiling fpring appear.

" The violet, that, thofe banks beneath,
    " Hides from thy fcorn its modeft head,
" Shall fill the air with fragrant breath,
    " When thou art in thy dufty bed.

" Even I, who boaft no golden fhade,
    " Am of no fhining tints poffeffed,
" When low thy lucid form is laid,
    " Shall bloom on many a lovely breaft.

" And he, whofe kind and foftering care
    " To thee, to me, our beings gave,
" Shall near his breaft my flow'rets wear,
    " And walk regardlefs o'er thy grave.

" Deluded flower, the friendly fcreen
  . " That hides thee from the noon-tide ray,
" And mocks thy paffion to be feen,
    " Prolongs thy tranfitory day.

" But kindly deeds with fcorn repaid,
 " No more by virtue need be done:
" I now withdraw my dufky fhade,
 " And yield thee to thy darling fun."

Fierce on the flower the fcorching beam
 With all its weight of glory fell;
The flower exulting caught the gleam,
 And lent its leaves a bolder fwell.

Expanded by the fearching fire,
 The curling leaves the breaft difclofed;
The mantling bloom was painted higher,
 And every latent charm expofed.

But when the fun was fliding low,
 And evening came with dews fo cold;
The wanton beauty ceafed to blow,
 And fought her bending leaves to fold.

Those leaves, alas! no more would close;
   Relaxed, exhausted, sickening, pale;
They left her to a parent's woes,
   And fled before the rising gale.

Pub. by A. S. Horn..g Post Mall.

FABLE IX.

THE BEE-FLOWER*.

COME, let us leave this painted plain;
 This wafte of flowers that palls the eye:
The walks of NATURE's wilder reign
 Shall pleafe in plainer majefty.

---

* This is a fpecies of the Orchis, which is found in the barren
and mountainous parts of Lincolnfhire,. Worcefterfhire, Kent, and

Thro' thofe fair fcenes, where yet fhe owes
　　Superior charms to BROCKMAN's art,
Where, crowned with elegant repofe,
　　He cherifhes the focial heart—

Thro' thofe fair fcenes, we'll wander wild,
　　And on yon paftured mountains reft ;
Come, brother dear ! come, Nature's child !
　　With all her fimple virtues bleft.

The fun far-feen on diftant towers,
　　And clouding-groves and peopled feas,
And ruins pale of princely bowers
　　On BEACHBOROUGH's airy heights fhall pleafe.

Hertfordfhire. Nature has formed a Bee apparently feeding on the
breaft of the flower with fo much exactnefs, that it is impoffible at
a very fmall diftance to diftinguifh the impofition. For this purpofe
fhe has obferved an œconomy different from what is found in moft
other flowers, and has laid the petals horizontally. The genus of the

Nor lifelefs there the lonely fcene ;
  The little labourer of the hive,
From flower to flower, from green to green,
  Murmurs, and makes the wild alive.

See, on that flow'ret's velvet breaft
  How clofe the bufy vagrant lies !
His thin-wrought plume, his downy breaft,
  Th' ambrofial gold that fwells his thighs !

Regardlefs, whilft we wander near,
  Thrifty of time, his tafk he plies ;
Or fees he no intruder near ?
  And reft in fleep his weary eyes ?

Orchis, or Satyrion, fhe feems profeffedly to have made ufe of for her
paintings, and on the different fpecies has drawn the perfect forms of
different infects, fuch as Bees, Flies, Butterflies, &c.

Perhaps his fragrant load may bind
   His limbs; we'll fet the captive free—
I fought the living Bee to find,
   And found the picture of a Bee..

Attentive to our trifling felves,
   From thence we plan the rule of all;
Thus NATURE with the fabled elves
   We rank, and thefe her *Sports* we call.

Be far, my friends, from you, from me,
   Th' unhallowed term, the thought profane,
That LIFE'S MAJESTIC SOURCE may be
   In idle Fancy's trifling vein.

Remember ftill, 'tis NATURE'S plan
   Religion in your love to find.;
And know, for this, fhe firft in man
   Infpired the imitative mind.

As confcious that affection grows,
    Pleafed with the pencil's mimic power*;
That power with leading hand fhe fhews,
    And paints a Bee upon a flower.

Mark, how that rooted mandrake wears
    His human feet, his human hands!
Oft, as his fhapely form he tears,
    Aghaft the frighted ploughman ftands.

See where, in yonder orient ftone,
    She feems ev'n with herfelf at ftrife,
While fairer from her hand is fhewn
    The pictured, than the native life.

---

* The well known fables of the Painter and the Statuary that fell
in love with objects of their own creation, plainly arofe from the
idea of that attachment, which follows the imitation of agreeable
objects, to the objects imitated.

HELVETIA's rocks, SABRINA's waves,
   Still many a shining pebble bear,
Where oft her studious hand engraves
   The perfect form, and leaves it there.

O long, my PAXTON*, boast her art;
   And long her laws of love fulfil:
To thee she gave her hand and heart,
   To thee, her kindness and her skill!

* An ingenious Portrait Painter in Rathbone Place.

Thos. Stothard del.   Pub. by A. & C. Hardwng, Pall Mall.   Arrnolt Sculp.

FABLE. X.

THE WILDING AND THE BROOM.

IN yonder green wood blows the Broom ;

Shepherds, we'll truft our flocks to ftray,

Court Nature in her fweeteft bloom,

And fteal from Care one fummer-day.

From Him * whofe gay and graceful brow
　　Fair-handed HUME with rofes binds,
We'll learn to breathe the tender vow,
　　Where flow the fairy FORTHA winds.

And oh! that He † whofe gentle breaft
　　In Nature's fofteft mould was made,
Who left her fmiling works impreft
　　In charaƈters that cannot fade.

That He might leave his lowly fhrine,
　　Tho' fofter there the Seafons fall—
They come, the fons of verfe divine,
　　They come to Fancy's magic call.

——————" What airy founds invite
" My fteps not unreluƈtant, from the depth
" Of SHENE's delightful groves? Repofing

• WILLIAM HAMILTON of Bangour.　† THOMSON.

" No more I hear the bufy voice of men

" Far-toiling o'er the globe—fave to the call

" Of foul-exalting poetry, the ear

" Of death denies attention.   Rouzed by her,

" The genius of fepulchral filence opes

" His drowfy cells, and yields us to the day.

" For thee, whofe hand, whatever paints the fpring,

" Or fwells on fummer's breaft, or loads the lap

" Of autumn, gathers heedful—Thee whofe rites

" At nature's fhrine with holy care are paid

" Daily and nightly, boughs of brighteft green,

" And every faireft rofe, the god of groves,

" The queen of flowers, fhall fweeter fave for thee.

" Yet not if beauty only claim thy lay,

" Tunefully trifling.   Fair philofophy,

" And nature's love, and every moral charm

" That leads in fweet captivity the mind

" To virtue—ever in thy neareft cares

" Be thefe, and animate thy living page

" With truth refiftlefs, beaming from the fource

" Of perfect light immortal—Vainly boasts
" That golden Broom its funny robe of flowers
" Fair are the funny flowers; but fading foon
" And fruitlefs, yield the forefter's regard
" To the well-loaded Wilding—Shepherd, there
" Behold the fate of fong, and lightly deem
" Of all but moral beauty."

———————" Not in vain"—
I hear my HAMILTON reply,
(The torch of fancy in his eye)
" 'Tis not in vain," I hear him fay,
" That nature paints her works fo gay ;
" For, fruitlefs tho' that fairy broom,
" Yet ftill we love her lavifh bloom.
" Cheered with that bloom, yon defart wild
" Its native horrors loft, and fmiled.
" And oft we mark her golden ray
" Along the dark wood fcatter day.

" Of moral ufes take the ftrife;

" Leave me the elegance of life.

" Whatever charms the ear or eye,

" All beauty and all harmony ;

" If fweet fenfations thefe produce,

" I know they have their moral ufe.

" I know that NATURE's charms can move

" The fprings that ftrike to VIRTUE's love."

FABLE XI.

THE MISLETOE AND THE PASSION FLOWER.

IN this dim cave a Druid sleeps,
　Where stops the passing gale to moan;
The rock he hollowed o'er him weeps,
　And cold drops wear the fretted stone.

In this dim cave, of different creed,
    An hermit's holy afhes reft :
The fchool-boy finds the frequent bead,
    Which many a formal matin bleft.

That truant time full well I know,
    When here I brought, in ftolen hour,
The Druid's magic Mifletoe,
    The holy hermit's Paffion-flower.

The offerings on the myftic ftone
    Penfive I laid, in thought profound,
When from the cave a deepening groan
    Iffued, and froze me to the ground.

I hear it ftill—Doft thou not hear ?
    Does not thy haunted fancy ftart ?
The-found ftill vibrates thro' mine ear—
    The horror rufhes on my heart.

Unlike to living founds it came,
  Unmixed, unmelodized with breath ;
But, grinding thro' fome fcrannel frame,
  Creaked from the bony lungs of death.

I hear it ftill—" Depart, " it cries ;
  " No tribute bear to fhades unbleft :
" Know, here a bloody Druid lies,
  " Who was not nurfed at Nature's breaft.

" Affociate he with dæmons dire,
  " O'er human victims held the knife,
" And pleafed to fee the babe expire,
  " Smiled grimly o'er its quivering life.

" Behold his crimfon ftreaming hand
  " Erect !—his dark, fixed, murderous eye !"
In the dim cave I faw him ftand ;
  And my heart died—I felt it die.

I fee him ftill—Doft thou not fee
   The haggard eye-ball's hollow glare?
And gleams of wild ferocity
   Dart thro' the fable fhade of hair?

What meagre form behind him moves,
   With eye that rues th' invading day;
And wrinkled afpect wan, that proves
   The mind to pale remorfe a prey?

What wretched—Hark!—the voice replies,
   " Boy, bear thefe idle honours hence!
" For here a guilty hermit lies,
   " Untrue to Nature, Virtue, Senfe.

" Tho' Nature lent him powers to aid
   " The moral caufe, the mutual weal;
" Thofe powers he funk in this dim fhade,
   " The defperate fuicide of zeal.

" Go, teach the drone of faintly haunts,
  " Whofe cell's the fepulchre of time;
" Tho' many a holy hymn he chaunts,
  " His life is one continued crime.

" And bear them hence, the plant, the flower;
  " No fymbols thofe of fyftems vain!
" They have the duties of their hour;
  " Some bird, fome infect to fuftain."

* 9 7 8 3 7 4 4 7 8 5 5 9 4 *